Cornwall Iron Furnace

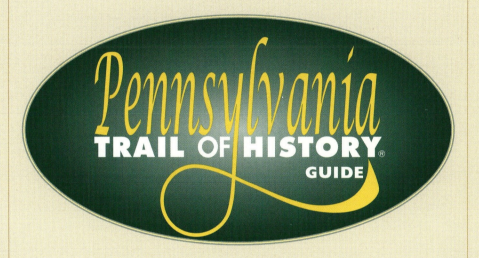

Text by Susan Dieffenbach
Photographs by Craig A. Benner

STACKPOLE BOOKS

PENNSYLVANIA HISTORICAL
AND MUSEUM COMMISSION

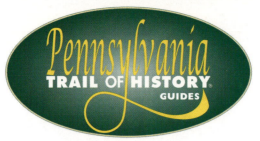

Kyle R. Weaver, Series Editor
Tracy Patterson, Designer

Published by
STACKPOLE BOOKS
5067 Ritter Road
Mechanicsburg, Pennsylvania 17055

Pennsylvania Trail of History® is a registered trademark of the Pennsylvania Historical and Museum Commission.

Printed in the United States of America
2 4 6 8 10 9 7 5 3 1
FIRST EDITION

Maps by Caroline Stover

Photography
Craig A. Benner: cover, 3, 5, 6, 11, 15, 16, 18, 20, 29, 31–47

Library of Congress Cataloging-in-Publication Data

Dieffenbach, Susan.
      Cornwall Iron Furnace : Pennsylvania trail of history guide / text by Susan Dieffenbach ; photographs by Craig A. Benner.—1st ed.
          p. cm.—(Pennsylvania trail of history guides)
      ISBN 0–8117–2624–X
      1. Cornwall Furnace (Cornwall, Pa.)—History. 2. Iron foundries—Pennsylvania—Cornwall—History. 3. Iron industry and trade—Pennsylvania—Cornwall—History. I. Title. II. Series.

TS229.5.U6 D54 2003
672.2'09748'19—dc21

                                                                              2002011182

Contents

Editor's Preface

For nearly two and a half centuries, iron manufacturing was a leading industry in Pennsylvania. Iron furnaces became prevalent in the early 1700s, and by the following century, the industry was the largest in the state. The Pennsylvania Historical and Museum Commission (PHMC) preserves the legacy of the state's early iron heritage at Cornwall Iron Furnace, which operated from 1742 until 1883. Stackpole Books is pleased to join with the PHMC in featuring the site in this new volume of the Pennsylvania Trail of History Guides, a series on the museums and historic sites administered by the PHMC.

The series was conceived and created by Stackpole Books with the cooperation of the PHMC's Division of Publications and Bureau of Historic Sites and Museums. Donna Williams heads the latter, and she and her staff of professionals review the text of each guidebook for accuracy and have made many valuable recommendations. Diane Reed, Chief of Publications, has facilitated relations between the PHMC and Stackpole from the project's inception, organized the review process with the commission, and attended to numerous details related to the venture.

The Administrator of Cornwall Iron Furnace, Stephen Somers, brought effective ideas and enthusiastic support to this particular book, and his recommendation for the author of the text was inspired. Back in 1999, when Steve was working at Ephrata Cloister, he suggested that I hire Craig A. Benner as a photographer for that site's volume. Since then, Craig has produced outstanding photography for several other books in the series, including this one.

Susan Dieffenbach, the author of the text, is a member of the board of directors of Cornwall Iron Furnace Associates and has been a volunteer tour guide at the furnace for sixteen years. On these pages, she explains the technology of ironmaking and its importance as an industry in Pennsylvania, recounts the human story of Cornwall Iron Furnace from its eighteenth-century origins to its preservation as a historic site, and guides the reader on a tour of the Furnace Building and other remaining structures on the site, which was once the nucleus of a great industrial plantation.

Kyle R. Weaver, Editor
Stackpole Books

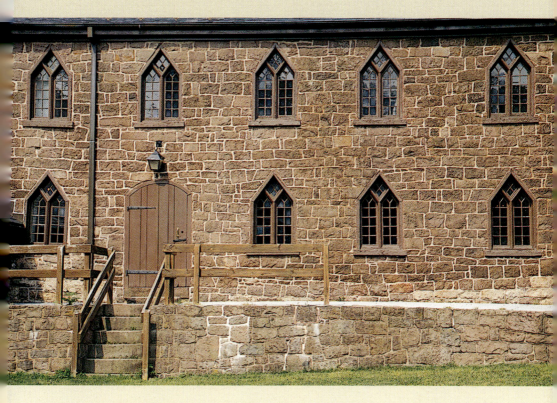

Cornwall Iron Furnace operated continuously from 1742 to 1883 and is typical of the charcoal, cold-blast iron furnaces that dotted the Pennsylvania landscape in the eighteenth and nineteenth centuries. As with other furnaces of its kind, all of the iron ore, wood for charcoal, limestone, and water power necessary for smelting iron were on hand in a favorable setting. Cornwall Iron Furnace is the only known surviving furnace of its kind in the Western Hemisphere. Its picturesque Gothic Revival buildings still stand near the remarkable world-class ore mine that operated from the 1730s until 1973. Today's Cornwall countryside, with its workers' villages, shops, schools, churches, and ironmaster's mansion, bears witness to the once-thriving iron plantation that had a flourishing iron furnace at its heart.

The five-acre site is a National Historic Landmark and a Landmark of the American Society of Mechanical Engineers. It includes remnants of the original eighteenth-century industry incorporated into the nineteenth-century Furnace Building, Wagon Shop and Blacksmith Shop buildings, Abattoir, and Charcoal Barn now refitted as a Visitor Center. In close proximity but not part of the site are the Ironmaster's Mansion, several workers' villages, and the Cornwall Iron Ore Mine, which today is filled with water. The site is administered by the Pennsylvania Historical and Museum Commission and is also supported by Cornwall Iron Furnace Associates, a volunteer group founded in 1984.

Ironmaking in Pennsylvania

We can scarcely imagine the value our ancestors placed on iron and iron products. Blacksmiths in the Middle East, using the most rudimentary equipment, labored to fashion wrought-iron products at least four thousand years ago. At the time of American settlement, most tools and many household items in Europe were made at least partially of iron. Emigrants to Colonial America sacrificed many comforts but were unwilling to do without iron. Native Americans produced no iron, and importing iron and iron products from Europe was prohibitively expensive, so early settlers had a powerful incentive to master ironmaking.

The techniques of ironmaking have been refined over many centuries, but the process has remained essentially unchanged: Intense heat is used to separate the iron in iron ore from impurities, the waste is discarded, and useful products are fabricated with the remaining iron. No matter the time period or technology, ironmaking is an arduous process. Feeding a blast furnace is like feeding a noisy, fire-breathing dragon, surrounded by flames, smoke, ashes, sparks, and soot, and whose main product by volume is carbon monoxide. Iron-making always involves locating large quantities of heavy rocks that must be dug, broken up, cleaned, hauled, and cooked. Enormous amounts of fuel are required to burn hot enough to melt the rocks, or at least soften them so the iron can be isolated by repeated hammering. Then limestone is needed for flux, which involves locating, digging, hauling, and cooking more rocks. Suitable equipment must be constructed to apply oxygen in order for an effectual fire to burn. Skilled and unskilled workers are needed in good number, and an appropriate location is required. Conditions in Colonial Pennsylvania were optimal for the development of a flourishing iron industry, and the commonwealth led the American colonies in their production of one-seventh of the world's iron in the mid-eighteenth century. The drama and events that shaped America into an industrial giant of the twentieth century were played out in the successful iron plantations of early Pennsylvania.

THE IRONMAKING PROCESS

The process of ironmaking consists of reducing the ore, separating the iron from any impurities that are present.

The Casting Room at Cornwall Iron Furnace.

The ancients knew how to create small tools and weapons by repeatedly heating ore-containing rocks in burning charcoal and pounding them until the impurities were driven out, the iron fibers were lengthened and toughened, and a small but useful piece of wrought iron remained. This process of direct reduction, removing oxygen from the iron, was accomplished in a bloomery, so named for the bloom, the mass of iron as it lay in the coals.

Bloomeries originated in the Middle East as small clay furnaces, manually heated by hand or foot bellows. After the thirteenth century, the application of water power allowed for larger bellows and more powerful hammers than hands or feet could operate, which increased the size of individual bloomeries and consequently the amount of finished wrought-iron product.

A new type of ironmaking developed in Europe a few centuries before the American colonies were settled. It used blast furnaces, which produced much larger quantities of cast iron by a method of smelting and indirect reduction in high heat. It is not necessary to melt the rocks to make wrought iron in a bloomery, but it is necessary to melt the rocks to make cast iron in a blast furnace.

The blast furnace so familiar to Colonial America was a flat-topped, truncated, hollow pyramid, usually twenty-five to thirty-five feet high and constructed in three layers: a stonework stack, a sandstone or firebrick lining, and a layer of rubble and other materials between the stack and the lining that provided insulation and allowed for expansion and contraction. Cornwall Iron Furnace was just one of many such blast furnaces that dotted the countryside of Colonial Pennsylvania, providing pig iron and useful products for residents of the commonwealth.

A blast furnace was fed, or charged, at somewhat regular intervals with charcoal, iron ore, and limestone placed into its mouth. Charging by buggy or basket continued day and night for as long as the furnace was in blast. The massive stonework of the furnace structure provided containment for an inferno of burning charcoal and functioned as support for the hollow, bottle-shaped lining of the furnace, where the iron was actually separated from the ore.

The shape of the furnace lining was critical to its operation. Downward from the mouth, the lining widened to about nine feet until it reached its center, the bosh, where the rocks would melt. A funnel was formed below the bosh, where the lining wall sloped inward to about four feet. This shape provided necessary support for the burning mixture so that the blast could pass through unhampered. Hot, molten rock material trickled down and was funneled into the crucible, where it was contained by specific density: heavy iron on the bottom, and impurities forming a layer on top. The crucible was emptied twice daily, yielding about two tons of cast iron at each tap.

Great consideration was given to the location and setting of the furnace. It needed to be located near the raw materials used in smelting, especially iron ore, which was the most difficult to find and the heaviest to transport. Access to a market for iron products was another factor. A good setting for a blast furnace needed to be dry, yet near a stream for water power, and the furnace had to be erected near a hill or bank to facilitate loading into the open top and emptying from the bottom. The success of ironmaking in Pennsylvania was expedited by favorable settings and the natural abundance of the ingredients for ironmaking: iron ore, fuel, and flux.

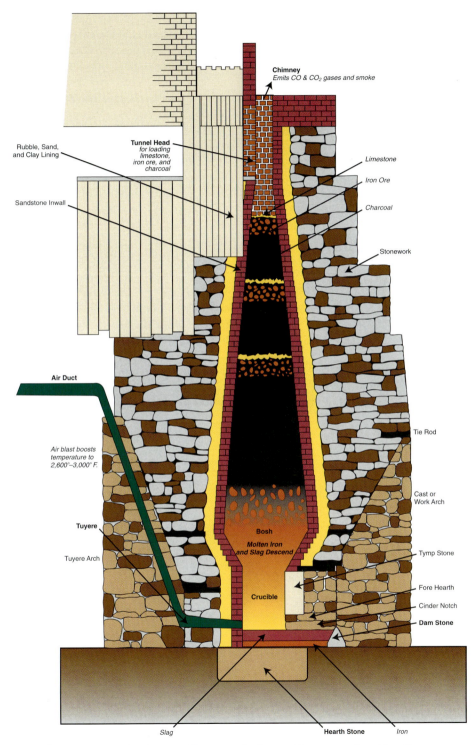

Chimney
Emits CO & CO₂ gases and smoke

Rubble, Sand, and Clay Lining

Tunnel Head
for loading limestone, iron ore, and charcoal

Sandstone Inwall

Limestone

Iron Ore

Charcoal

Stonework

Air Duct

Air blast boosts temperature to 2,600°–3,000° F.

Tie Rod

Cast or Work Arch

Tuyere

Tuyere Arch

Bosh
Molten Iron and Slag Descend

Tymp Stone

Fore Hearth

Cinder Notch

Dam Stone

Crucible

Slag

Hearth Stone

Iron

***Manufactuing Iron.** This diagram illustrates the ironmaking process in a blast furnace.*

IRON ORE AND IRON

Iron (Fe) is a metallic element with a melting point of about 2100 degrees F, the fourth most widely distributed element in the earth's crust, and, after aluminum, the second most widely distributed metallic element, making up about 5 percent of the earth's crust. In nature, iron is found as iron ore combined with oxygen, sulfur, silica, and other elements. There are many kinds of iron ore, including red and brown hematite, limonite, taconite, peronite, siderite, carbonate, and bog ore, in varying degrees of richness. Magnetite ore, quite rich (over 60 percent iron) and lower in oxide than other ores, was located at Cornwall in three hills—later named Big Hill, Middle Hill, and Grassy Hill—collectively known as the Cornwall Ore Banks. This ultimately represented one of the most valuable magnetite ore deposits in the eastern United States. The mine was also a mineralogist's paradise, with recoverable amounts of copper, gold, silver, and cobalt; eighty-six minerals; and a remarkable pocket of zeolite.

Although iron can take many forms, because of its ability to chemically unite with other elements, it can be classified into three major groups: wrought iron, cast iron, and steel. Wrought iron, used for horseshoes, tools, and nails, is the oldest form, dating back at least four thousand years. It is chemically pure, strong in tension, with a sinewy, fibrous structure, and it can be shaped by hammering (by the blacksmith, for example), squeezing, or rolling. Cast iron, such as that used to make frying pans, is an alloy of iron and carbon. The carbon

North Cornwall Furnace, 1902. Solid iron was removed from mines by blasting with dynamite. CORNWALL IRON FURNACE

The magnetite ore mined in the early days at Cornwall was of high quality, sometimes over 60 percent iron. As mining continued, the grade of the ore that was obtained dropped. The amounts of sulfur and copper in the ore increased, and it became apparent as early as 1883 that some form of ore concentration would have to be done in order to successfully compete with ores from other areas.

Sulfur could be removed by weathering the ore over time or by roasting, first on a pile of burning wood, later in roasting ovens, and finally by sintering. Magnetic separation of the iron from other minerals provided an early solution, as in the magnetic ore-separating experiments of Thomas Edison. In 1905, iron was magnetically separated on a commercial scale, and by 1916, copper and magnesium were concentrated and separated from the ore using vacuum and oil flotation methods. Cornwall became America's leading supplier of cobalt during World War II, when further experiments accomplished the separation of cobalt from pyrite, an iron-containing mineral.

Experiments in Lebanon with pelletizing resulted in the country's first successful pellet furnace and pellet plant in 1950, replaced in 1962 by a modern plant erected at the open-pit mine at Cornwall. Pelletizing is a way of concentrating iron in ore by crushing the ore, mixing it with a binder, and baking it into balls. This represented the last application of science to keeping the ore commercially competitive and the mine profitable.

in cast iron imparts a crystalline structure, which makes it weak in tension and impossible to hammer into shape. Thus it must be shaped into pig iron for further refining or into finished products by molding. Steel is the most versatile and widely used form of iron and can have a simple chemical composition or be a complex alloy. Neither wrought iron nor steel were produced at blast furnaces, such as Cornwall Iron Furnace.

Cast Iron Products, c. 1800.

FUEL

Although British ironmasters had switched to using anthracite coal for fuel as early as 1750, mostly as a result of deforestation, charcoal remained the main fuel in Colonial America, with the charcoal cold-blast technology dominating American furnaces, until the mid-nineteenth century. Charcoal is wood made into almost pure carbon, which burns in a blast furnace at a temperature between 2650 and 3200 degrees F, hot enough to melt rocks. The carbon from the charcoal also becomes part of the iron fabric, giving cast iron its characteristic color, short fiber, and brittle quality. Each day that a furnace was in blast, it consumed an entire acre of trees in the form of charcoal. The production of the mass quantities of charcoal required by the furnace was an industry in itself,

CHARCOAL MAKING

Wood was coaled, or made into charcoal, by slowly roasting it under carefully controlled conditions in hearths, or pits, thirty to forty feet in diameter. The process begins with cutting and stacking wood, which occurred when farmers and farm laborers were available to help—late fall, winter, and early spring. Woodcutters comprised the largest portion of the work force at the iron plantation. (Peter Boyer, son of a Cornwall collier, wrote that each woodcutter worked in a seventy-five foot swath.) Hard woods such as chestnut, hickory, oak, beech, and walnut were preferred because they yielded high quality charcoal, but soft woods, such as pine, were sometimes used.

The wood was hauled and stacked at dry, level coaling sites, free from stones but not sandy or loamy, and sheltered from the wind. Between May and November, the highly skilled collier and his helpers carefully stacked about twenty-five to fifty cords of the prepared wood (billets and lapwood) on end in layers around a six to eight foot chimney of sticks. This mound was then covered with leaves and dirt to restrict the flow of air, with strategically placed air holes, and set on fire by lighting kindling in the chimney. The flame was smothered when it was well underway to create a smoldering, smoky mass, having just enough heat to drive out all the water, tar, and other substances from the wood and leaving behind nearly pure carbon. An untended mound could easily become a raging bonfire, resulting in ashes rather than charcoal. The collier lived in the woods and carefully tended the pit day and night for ten to fourteen days until it was completely charred. To prevent gas pockets, workers tramped the mound several times each day, locating and refilling soft spots as necessary. When the collier judged the coaling was finished, the fire was extinguished by choking off all the air. The outer covering was removed after the mound cooled for ten or twelve more days. Workers raked out the pile into smaller, manageable rings, using dirt to put out new fires as oxygen was introduced, and when these were sufficiently cool, teamsters hauled the charcoal to the furnace site where it was stored for imminent use. Wet charcoal is unus-

Transporting Wood for Charcoal. Vast amounts of wood were needed for making charcoal, an essential ingredient in cast iron. PENNSYLVANIA STATE ARCHIVES

with its own hierarchy of skilled colliers, who directed the burning of the wood to make charcoal; teamsters, who hauled the charcoal to the furnace site; and unskilled laborers and wood cutters. Many changes in the labor force resulted when ironmasters eventually responded to advances in technology and switched to anthracite coal or bituminous coke.

FLUX AND SLAG

Limestone, rich in calcium carbonate, was also added to the furnace, where it worked effectively in the high heat as a flux to remove impurities both chemically and by physically sticking to the dirt and stones in the ore, known as gangue. The substance formed by the combination of the limestone with

Colliers tended the mounds of wood that smoldered into charcoal. PENNSYLVANIA STATE ARCHIVES

able, which made large charcoal barns, sheds, or houses very important buildings at an iron site. Transportation and storage problems arose as nearby forests were cleared and the work was carried on at ever greater distances from the plantation. Some later furnaces made charcoal in onsite charcoal ovens that facilitated the extraction of other chemicals.

The collier's life and making charcoal were woven inextricably together. From the time the pit was first fired in May until teamsters hauled out the last piece of charcoal in late October, he lived in the woods, constantly tending several hearths he and his helpers coaled simultaneously. Maintaining the right amount of oxygen in the pile was a delicate balancing act that left little time for food gathering—meals

were often bread and coffee. Serious injury, illness, and economic or physical disaster were constant perils. The collier's home was typically a crude, conical pole hut, about ten feet high and eight feet across, covered over with leaves and dirt and furnished with a one-man door, a wood stove, and log bunks.

Many colliers were regarded as hermits, while other more colorful characters were the subject of local legend. One of the colliers working at Cornwall Furnace was an African American called Governor Dick, whose home was a cabin at the foot of the hill east of Mount Gretna, a wooded area west of the furnace. The hill was known for some time as Governor Dick's Hill, but has since become known simply as Governor Dick.

the impurities in the iron ore was called slag. The molten slag trickled into the crucible along with the molten iron, and being much lighter and less dense, it formed a layer over the heavy iron, much like the fat rising to the top of gravy. The layer of slag helped retain heat in the iron until it was drained off, taking cinders and impurities with it.

Conditions in the furnace and the quality of the final iron product could be read from the color and texture of the slag—for example, protoxides of iron produced shiny black and green "furnace glass" slag. Along with iron ore and trees for charcoal, beds of limestone were abundant and easily quarried in Pennsylvania.

OXYGEN

To maintain intense furnace temperatures hot enough to melt rocks, it was imperative to establish a steady stream of compressed air, or blast. As the blast bubbled up through the burning mass in the furnace, it fed oxygen to the fire, hastening the separation of the iron from the ore. The first universal blast equipment was a water-powered bellows that compressed the air and then fed it to the furnace through a pipe called a tuyere. Soon pairs of closed blowing cylinders or blowing tubs, still powered by water wheels, began to replace the old leather bellows. The water wheel drove pistons up and down reciprocally in the tubs, which were lined with leather to make them airtight. By a simple system of intake and exhaust valves, air from the tubs was forced into a central mixing box and then through a duct to the tuyere. Blast operations became even more efficient in the nineteenth century as water turbines and steam engines replaced the water wheels, which enabled many furnaces to increase the number of tuyeres. The steam engine at Cornwall was most likely installed in 1841. It was not too long before hot air was found to be even more effective at reducing iron ore. Some operations refitted their charcoal cold-blast furnaces for hot-blast operation, while other ironmasters replaced or rebuilt their furnaces to accommodate the modern anthracite hot-blast technology.

FURNACE OPERATIONS

It took two days to blow in, or start, a new furnace or to restart a furnace that had blown out for repair or annual cleaning, a necessary procedure that took six weeks to two months. Fillers loaded charcoal into the tunnel head, or mouth of the furnace, until the furnace was completely full. The charcoal was set on fire at the top and allowed to burn all the way to the bottom. The fillers then fed the furnace with charcoal until it was full again, this time burning from the bottom until the stack was hot and nothing but glowing coals remained.

Then it was time to charge the furnace, adding alternate layers of charcoal, iron ore, and limestone to the fiery mass at regular intervals. Fillers charged the furnace day and night with the proportions of these three materials as determined by the founder based on his basic analysis of the ore on hand. Charging continued day and night for as long as the furnace was in blast, eighteen to twenty charges a day. As the temperature approached 2500 to 3200 degrees F in the bosh, the rocks began to melt and the furnace to work. Slag was removed and discarded every hour or half hour. In this continuous process, it took about forty hours for a batch of iron ore to go from mouth to hearth.

Twice a day, the furnace was tapped. At about noon and midnight, at the founder's signal, workers gathered in the casting house at the bottom of the furnace stack. The keeper unplugged the cinder notch, the top hole in the dam stone or door, which allowed the remaining liquid slag to drain off to one side. After the slag was cooled, workers broke it up and discarded it as waste. The lower plug in the dam stone was then removed, and red hot, molten iron flowed through channels into the molds already prepared by the molders in the sand floor of the cast house. Guttermen guided the flow from the large channel into the smaller molds formed at right angles. This figure so resembled a sow with nursing piglets that the main product of the blast furnace became known as pig iron. Pigs, the first step in extracting iron from ore, had a flat side and could be stored or transported to market or to the forge for further refining.

A Pig Bearing the Grubb Imprint.

Some of the liquid iron was also cast as ironware by molders, who ladled molten iron from the furnace into special casting flasks to produce common items such as skillets, kettles, and stove plates. Using skillfully carved molds, molders often cast stove plates with intricate designs that included the furnace and ironmaster's name, date of casting, Bible verses, and other mottoes and symbols.

THE REFINERY FORGE

At the refinery forge, the pig iron was heated and worked in a charcoal fire to remove cinders, carbon, and slag. This pasty loupe was then repeatedly heated and pounded under a water-driven hammer to remove more cinders and produce bars with the long, tough fibers of wrought iron. Blacksmiths used these merchant bars to fabricate the horseshoes, nails, weapons, wagon parts, gates, tools, and other items so necessary to everyday life. A rolling and slitting mill could further transform wrought iron into thin slit iron, from which nails were cut.

STEEL

Steel began to challenge the dominance of iron soon after the close of the American Civil War. A tiny but critical percentage of carbon was alloyed with iron for use in products that had to withstand abnormal wear or flexing. The early process was cementation, in which bars of wrought iron were kept at a red heat for up to two weeks in a pot packed with charcoal or charcoal dust. The bars absorbed about 1 percent of their weight as carbon and were converted to blister steel, which had a characteristic blistered surface. These glowing bars or strips of steel were hammered, reheated, and rehammered several times to drive out slag, which produced a scarce, expensive, and high grade of steel.

Ancient cultures had used steel to fashion superior weapons; in Colonial America, thin strips of steel were welded onto axe blades and plow moldboards. Eventually the demand for steel led to technological advances, from the crucible method, to the Bessemer converter, and finally, mechanized Gilchrist-Thomas open-hearth furnaces, which produced affordable mass quantities of good steel. The railroad industry's sustained demand for steel rails resulted in heretofore unseen developments in methods, labor, and economy. By 1883, the annual output of steel in the United States equaled that of iron, and it was not long before iron was superseded by steel as the dominant industry.

The History of Cornwall Iron Furnace

The years 1715 to 1776 saw an abundance of new iron works in Colonial Pennsylvania. The construction in that period of at least twenty-one blast furnaces (including Cornwall Iron Furnace), forty-five forges (including Upper and Lower Hopewell Forges), four bloomeries, six steel furnaces, three slitting mills, two plate mills, and one wire mill attest to the tremendous demand for iron products in the new nation. The history of Cornwall Iron Furnace is a tale of initiative and enterprise, of force of character and unusual executive ability in the American tradition. These traits were manifested at Cornwall in two families: the Grubbs and the Colemans.

THE GRUBBS

The first ironmaster of Cornwall Furnace was Peter Grubb. The Grubb family had come from Cornwall, England, in the late seventeenth century and established themselves along the Delaware coast, where John Grubb, Peter's father, had a tannery. Later the family moved to Marcus Hook in Chester County, Pennsylvania. Here, in 1700, Peter was born, the youngest of nine children.

After a successful apprenticeship as a stone mason, Peter converted to Quakerism in order to marry Quakeress Martha Wall in 1732. Their first son, Curttis, was born in 1733. It was not long afterward that Peter moved to Lancaster County, seeking good locations for stone quarrying. There, next to his own land, he found iron ore outcroppings.

By 1739, Peter's enterprises were well under way: He had accumulated over a thousand acres of property, mining operations were in progress, an indenture was made for the construction of a charcoal furnace to be called Cornwall, and construction on two Hopewell Forges on Hammer Creek had begun. His Hopewell Mansion, which still stands on Route 322, was probably substantially constructed by this time. In 1740, Martha died while giving birth to a second son, named after his father, and a little over a year later, Peter married his second wife, Hannah Mendenhall Marshall.

This Massive Wheel at Cornwall Furnace was run by a steam engine and provided the power to blast the great volumes of air required to melt the iron ore, charcoal, and limestone to make cast iron.

The furnace at Cornwall was in operation by 1742. In 1744, Grubb placed a notice in the *Pennsylvania Gazette* offering to lease Cornwall Furnace:

To be LET, A Furnace, Sawmill and Forge, within 13 Miles of the City of Lancaster, for 20 years, or otherwise as may be agreed upon giving good Security if required; all of them being almost new, with good Water and Timber, with an unquestionable Quantity of good Iron Ore, laying near so that three men and two Horse has and can supply her with Ore every Day, when she is in Blast. Pigmetal is proved to be very good, Hearth-stones handy, Limestone Sand and Twere Clay on the Premises, a Quantity of Coals housed, some Wood cut: She may be put into Blast early this Summer if required. There are 80 Acres of Land within Fence, 20 Acres of Meadow cleared, 50 more may be easily made, 7000 Rails ready mauled, with other Conveniences, to accommodate an Iron-Work. If any Person hath a Mind to lease the said Works, let him or them repair to the said Place and treat with Peter Grubb on Conditions.

The Cornwall Company, a group of twelve Quaker businessmen led by Amos Garrett, was granted a twenty-year lease in 1745 on the furnace, producing stove plates as well as pig iron. The lease eventually was held by only two men, Amos Garrett and Jacob Giles of Maryland.

The Grubb family moved to Wilmington, Delaware, in 1745, where Peter bought and sold real estate until his death in 1754. Peter Grubb left no will, and under the intestate laws of the day, his elder son, Curttis, received two-thirds title to the property, and his younger son, Peter, one-third. The gregarious Curttis Grubb was married and had a son, whom he named Peter Grubb Jr. (The designation Jr. was used historically to differentiate between men, not necessarily father and son, living in proximity.) Curttis was soon separated from his wife. He sent Peter Jr. to live with relatives and spent six years abroad before returning to Hopewell Forge to take up management of the iron operations. His brother Peter, on the other hand, was somber and ambitious, and desiring to succeed, learned financial matters from his uncle and guardian, Samuel Grubb. Youthful employment at Sarum Forge helped him to also learn ironworking. He took up residence at Hopewell Mansion in 1762, anticipating the expiration of the lease, to work with Jacob Giles in the Upper and Lower Hopewell Forges. When the Cornwall Furnace lease expired in 1765, the two brothers took over management of the property, residing together at Hopewell Mansion. Peter continued operation of the smaller but more lucrative Upper and Lower Hopewell Forges, and Curttis, who had less expertise, took over the larger but less profitable Cornwall Furnace.

Curttis remarried in 1773 and moved his family, including son Peter Jr., to his newly constructed ironmaster's mansion very close to Cornwall Furnace. In 1771, at age thirty-one, Peter had married eighteen-year-old Mary Burd Shippen, with whom he had two sons, Alan Burd Grubb (1772) and Henry

A Stove Plate, created at Cornwall Furnace, bearing the name of Curttis Grubb.

Bates Grubb (1774). Peter and his family lived at Hopewell Mansion. Mary died unexpectedly following the birth of Henry; Peter never remarried.

Now the lives of Curttis and Peter Grubb were swept up in the events surrounding the American Revolution. The brothers supported the patriot cause, both serving actively as colonels, commanding local units of Pennsylvania militia. Peter Jr. also joined the Continental Army and eventually spent seven months as a British prisoner of war after his capture at Fort Mercer. Cornwall lands served to pasture Army horses, and the manager at Cornwall Furnace cast forty-two cannons, piles of shot, cannonballs, salt pans (used in making salt peter for gunpowder), and stoves for the Continental Congress. The heavy cast-iron cannons were transported to Philadelphia to be put on board Naval frigates. Shipments of Colonial pig iron and unfinished bar iron across the Atlantic had increased after British Parliament passed the Iron Act in 1750. Stipulations of the Iron Act prohibited new Colonial finishing mills and steel furnaces in order to encourage export of unfinished Colonial iron to England, where decades of charcoal making had depleted the forests. The domestic demand for iron at this time was very great, and inflation during the five years of the war caused iron prices to nearly double. The Grubbs and other Pennsylvania ironmasters stood to realize handsome profits while in the service of their new nation.

After the Revolutionary War, the Cornwall property began to pass out of the hands of the Grubb family. The relationship between Curttis and Peter Grubb, once informal and cordial, became strained when Curttis announced plans to marry a third time. Curttis and Peter agreed to give Peter Jr.

CHRONOLOGY

c.1734	Peter Grubb, a Chester County stone mason in search of building materials, discovered extensive mineral deposits in northern Lancaster County
1737	Peter Grubb granted final warrant on Middle Hill, Big Hill, and Grassy Hill, a three-hundred-acre tract, and builds a bloomery to test the value of his ore
1739	Mining operations initiated
1742	Cornwall Iron Furnace begins operation; stone mason Peter Grubb is now ironmaster
1754	Peter Grubb dies, passing the operations at Cornwall to sons Curttis and Peter
1776	Revolutionary War engages all the colonies and elevates the Colonial iron industry to greater importance
1784	After disputes, Peter builds a furnace and residence south of Cornwall, naming it Mount Hope
1786–98	Robert Coleman becomes ironmaster of Cornwall Iron Furnace, the first of four generations of Colemans who will oversee Cornwall
c.1800	Blowing tubs replace bellows for blast at furnace
1825	Robert Coleman dies, and his interests in Cornwall go to his four sons: William, James, Edward, and Thomas Burd
c.1841	Steam engine arrives at furnace to power blowing tubs

(continued)

Revolutionary War Cannonballs manufactured at Cornwall Furnace.

his inheritance right away. It was therefore necessary to define more precisely the distribution of the Grubb assets, and the assessments and property partitions involved in formal estate planning caused unresolved tensions among them. In 1783, Curttis granted a one-sixth interest in Cornwall Furnace and one-third of the Hopewell Forges to Peter Jr. Thus began a fractionalizing process of the Cornwall ore body that was to become so complicated that it eventually led to a ninety-six-part ownership. Three years later, Peter Jr. sold his interest in Cornwall to Robert Coleman, reserving for himself and his heirs the right to obtain sufficient ore to supply one furnace for "as long as grass grows and water flows." This right later became vested in what was to be known as the Robesonia Iron Company, the last holdout to conglomerization in the twentieth century by Bethlehem Steel Corporation. Peter Jr. moved his family to Philadelphia after selling his interest in the mine to George Ege of Robesonia Furnace.

Meanwhile, in 1784, the elder Peter purchased approximately 212 acres to the south of Cornwall and built a furnace and mansion there, naming it Mount

Hope. It was probably his intention to continue operating the Hopewell Forges with pig iron smelted from Cornwall iron ore at his new Mount Hope Furnace, rather than buy Cornwall Furnace pig iron from his brother. Peter sought court action, but a decision was delayed because there seemed to be no equitable way to divide the Cornwall and Hopewell properties. Since no one knew the extent of the ore deposit, one or the other would be deprived of ore. Curttis allied with Robert Coleman, and in December 1785, Peter reluctantly agreed to partition the properties. Peter had no peace over this agreement. He became increasingly touchy and irritable, drinking more heavily, and eventually committed suicide on January 17, 1786. Peter's death dissolved any prior agreements on partitioning, and the matter was turned over to the courts. After deliberating for over a year, the jury still could not come to a fair division, but its recommendations were mutually agreeable to all parties: Curttis Grubb and Robert Coleman were to receive the Cornwall Iron Furnace and 6,520 acres of land, and Peter's sons, Alan and Henry, would own the Hopewell Forges and 3,741 acres of land. Each would also receive a share of the ore banks, which would remain undivided. Alan, having little interest in the iron business, sold his share in his father's estate to Henry, studied medicine, became a doctor, and moved to Tennessee. Henry successfully operated the Mount Hope Furnace and established a dynasty of wealthy nineteenth-century ironmasters in Lancaster County. Curttis Grubb died in 1789, passing his estate to his son Curtis Jr., who died in 1790 at age seventeen without an heir. His sister Elizabeth received the property, which she sold to Robert Coleman on her twenty-first birthday in 1795.

THE COLEMANS

Robert Coleman was born in Castlefin, Ireland, in 1748. Opportunities for economic success were limited in Ireland at this time, even to those of genteel birth, so Robert emigrated to Pennsylvania at age sixteen. In two years, he rose from a prothonotary's clerkship in Philadelphia to bookkeeper for Curttis and Peter Grubb, from whom he learned a great deal about ironmaking.

Serving next as clerk for ironmaster James Old at Quittapahilla Forge in Lebanon County, Coleman married Old's daughter Ann in 1773, the same year he leased Salford Forge near Norristown, Montgomery County. With the help of his father-in-law, he was able to make timely investments in the iron industry during the years before and after the Revolutionary War, a period of great demand for iron. He leased Elizabeth Furnace in Lancaster County, where he manufactured cannonballs and shot. With his wartime profits, he acquired a two-thirds share of Elizabeth Furnace, purchased Speedwell Forge, and bought into Cornwall and the Upper and Lower Hopewell Forges. In 1791, he built Colebrook Furnace. Three years later, he purchased the balance of Elizabeth Furnace; half of Henry Bates Grubb's share of Cornwall and the Hopewell Forges; and all of Curttis Grubb's share of the Cornwall Ore Banks and the Cornwall Iron Furnace. By 1798, he owned all of the shares of Cornwall Furnace, 5/6 of the mine, and by 1803, the Hopewell Forges, Union Forge, Martic Forge, and Speedwell Forge.

By the age of fifty-five, Robert Coleman had built a vast and complete iron empire, comprising the ore-rich mine, the furnace for smelting ironware and pig iron, several forges for refining pig iron, and almost ten thousand acres of

CHRONOLOGY

1849	Cornwall Anthracite Furnace built by Robert W. and William Coleman near the present-day community of Anthracite (Goosetown)
1861–65	Civil War increases demand for iron
1864	Cornwall Ore Bank Company established to regulate mining
1872	Burd Coleman Furnace built by heirs of Robert W. Coleman to use modern anthracite, hot-blast technology (remodeled in 1880)
1872–73	North Cornwall Anthracite Furnace constructed by Margaret Coleman Freeman (remodeled in 1887)
1883	Cornwall Iron Furnace blown out for the last time, after 141 years of continuous operation
1890	Coleman Anthracite Furnace razed
1916	Bethlehem Steel Company enters the Cornwall scene
1922	Burd Coleman Furnace dismantled
1923	North Cornwall Anthracite Furnace dismantled
1932	Margaret Coleman Freeman Buckingham presents the furnace property to the Commonwealth of Pennsylvania for a public museum
1973	The mine closes, and mining at Cornwall comes to an end after 234 uninterrupted years of production

Robert Coleman. *The shrewd Irish businessman who built an iron empire in Pennsylvania was painted by Lancaster artist Jacob Eichholtz (1776–1842) around 1820.* ©BOARD OF

land for charcoal, water power, farms, and workers' homes. His consistently wise investments enabled him to build an iron dynasty that lasted for four generations.

But wealth and position did not protect the Coleman family from tragedy. Robert lived in the mansion at Elizabeth Furnace until 1809, when the prosperous Coleman family moved to a townhouse in Lancaster. By this time, sons William, James, Edward, and Thomas Burd were well established on their own, and daughters Margaret, Harriet, and Elizabeth soon married. Another daughter and son died in the following two years, leaving two more daughters, Ann Caroline and Sarah, at

CORNWALL'S WORKERS

Cornwall Furnace was at the heart of a small, self-contained community centered around the furnace and the ironmaster's mansion. Here lived the workers and families who supported the furnace operations and were in turn sustained by its prosperity. The demands of the furnace—filling and tapping—established the rhythms of everyday life for the hundred or more workers who mined and smelted the iron ore; cut the wood; grew, harvested, and processed the food; cared for the animals; sewed and laundered the clothing; and tended the children.

The entire operation of the furnace was owned and directed by the ironmaster, who needed to have technical knowledge as well as business acumen and skill in labor relations. The ironmaster was assisted and advised by the clerk, a second in command, who kept the records, paid the bills, ordered supplies, and managed the company store. The responsibility to produce good iron fell to the founder, who carefully monitored the temperature in the stack, adjusting the fuel or the blast as necessary, and determined the proportions of ore, limestone, and charcoal to be fed into the furnace. The founder supervised the fillers, whose job it was to feed the furnace, and called for the tap when the iron and the molders were ready. A keeper assisted the founder, minding the furnace when the founder was away. Molders were skilled workers in the casting house who prepared the casting floor or flasks to receive the hot iron at tapping. Other workers in the community included race cleaners, planters, reapers, pickers, corn huskers, flax pullers, mowers, haymakers, dung spreaders, cider makers, planters, sheep shearers, and meat hangers.

Most of the managers, skilled workers, and some unskilled workers were free laborers who had immigrated from England. Workers in the unskilled positions were more likely to be indentured servants from Germany, England, and Ireland who had agreed to work for room, board, and clothing for a certain number of years in exchange for passage to America. Indentured servants often proved problematic and were likely to run away. Some of these

Miners Village, one of several workers communities that sprang up near the iron furnace.
CORNWALL IRON FURNACE

runaway indentured servants are described in advertisements for their return: "He is no scholar and pretends to be a Blacksmith." "He is a fair spoken talkative Fellow, when in Drink, which he is very much addicted to."

Like many other furnaces in the area, Cornwall Furnace also used slave labor for a time. Slaves worked at both skilled jobs, such as founders and colliers, and unskilled, such as woodcutters, teamsters, and miners. The number of slaves working at Cornwall began to decline in 1790, and the last slave appears in the records in 1798.

During the intense labor shortage caused by the Revolutionary War, Hessian prisoners of war, German mercenaries who had been captured while being paid to fight for the British, were permitted to work at Cornwall Furnace. When the war ended, a few decided to seek opportunities in the new nation and remained in the area.

There was a great disparity in living conditions between the workers and the owners. The ironmaster lived with his family in a mansion on a large estate with gardeners and servants, much like English gentry. Workers lived at the opposite end of the social and economic spectrum but were well treated. The Coleman family of ironmasters brought preachers and teachers to their workers, built churches, and gave out turkeys at Christmas.

Over the years, several villages were built by the ironmaster for the workers, and these communities are now incorporated into Cornwall Borough: Miners Village; Anthracite (known as Goosetown); Burd Coleman, and North Cornwall, in proximity to the twin anthracite furnaces. Rexmont was built by mine employees who wanted their own homes. Toytown, the center of Cornwall, was built in the 1950s by Bethlehem Steel Corporation for its workers.

home. It was in Lancaster, sometime during 1818, that daughter Ann Caroline, unmarried at twenty-two, began seeing James Buchanan, a tall, handsome lawyer employed by Molton Rogers. Despite her parents' doubts about his background and motives, Ann Caroline and Buchanan were soon engaged. Robert Coleman was familiar with Buchanan's academic and disciplinary problems at Dickinson College (he had been dismissed, reinstated, and come under faculty discipline) and was wary of anyone he believed might have designs on his wealth. It seems that Ann also grew suspicious of Buchanan's motives, especially when his work kept them apart more and more. She broke the engagement after an innocent but upsetting incident in which Buchanan tarried at the home of an acquaintance before returning to Ann after a trip to Philadelphia. To help her recover from her depression, Robert sent her to Philadelphia, where she was found dead from "hysterics" just after midnight on December 9, 1819. Most people suspected suicide, and Buchanan was devastated. A portrait of Ann Caroline

Coleman still hangs at Wheatland, Buchanan's residence in Lancaster, and many believe he never married because of his failed relationship with her.

Ann Caroline's sister Sarah also met a sad end. After her father's death in August 1825, Sarah hoped to marry William August Muhlenberg, co-rector of St. James Episcopal Church in Lancaster. Robert and Muhlenberg had been involved in a bitter dispute over evening worship services, thus preventing Sarah's engagement while he lived. But in his will, Robert Coleman had granted his sons Edward and James the right of approval of Sarah's future husband. Edward disliked Muhlenberg and forbade the engagement. Sarah fled to Philadelphia, where she took her life.

When Robert Coleman died in 1825, his Cornwall properties were willed to his four surviving sons. Two of the brothers, William and Edward, sold their interests to the others, James and Thomas Burd. Thomas Burd Coleman ran Cornwall, while James lived at Elizabeth Furnace. After James died at age forty-seven in 1831, Thomas Burd sued James's heirs, Robert and

Gothic Revival features were added to the buildings at Cornwall Furnace during a mid-nineteenth-century remodeling project. This view of the Furnace Building was photographed around 1860. CORNWALL IRON FURNACE

THE CORNWALL ORE MINES

The story of Cornwall as an iron plantation and community begins with its ore mine. This seemingly inexhaustible reserve became the largest open-pit iron ore mine in the world before the ore-rich Lake Superior region was developed.

There were two major ore bodies at Cornwall: the original outcropped discovery of Peter Grubb, and a second deposit located in 1919, farther east and 150 feet below the surface. Mining the outcropping began in the 1730s with picks and shovels, using wheelbarrows to move the ore. By 1848, after 106 years of continuous surface mining, an estimated 776,000 tons of ore had been removed.

Mining methods began to change in this open-pit mine, as the random holes were replaced in 1858 by terraces, or wheel and bench mining, which consisted of working the ore in benches, or steps, sixteen to twenty feet deep, still hauling by wheelbarrow to cars. Improvements to the mining operation quickly accelerated to meet the demands of industry. In the 1870s, steam shovels loaded the ore into railroad cars, which were hauled out by steam locomotives. In 1910, a 45-degree skip hoisting incline eliminated the long, steep rail haul, and a screening and crushing plant was added to replace hand sorting. In 1916, electric shovels replaced the steam shovels. And in 1944, the steam locomotives were replaced by diesel side-dump trucks.

Underground mining and open-pit mining were carried on simultaneously at Cornwall from 1921 until 1940. An incline shaft was sunk near the west end of the open pit, and drilling and blasting methods were employed until 1940, when this mine was closed so as not to disturb the ore still available from the cheaper open-pit mining. By 1953, all the available exposed ore had been removed, and it was necessary to convert to underground methods entirely. This ended an operation that had produced more than 65 million tons of iron ore and made Pennsylvania the leading iron-producing state for many years. When the open pit closed, geologists estimated that a twenty-year supply of ore remained, and underground operations resumed successfully, using block caving, in which the ore body is weakened and falls into a prepared haulage way. A single mine shaft hoisted men, material, and ore.

Development of the second ore body began in 1926, but it was curtailed in 1931. Underground mining resumed here in 1936, using two parallel shafts and a block caving method, eventually reaching the depth level of the ore at 1,225 feet. Other improvements, including slusher drifts (harvesting ore by lique-

George Dawson Coleman, for a more equitable distribution of their father's iron empire. As a result, James's family got Elizabeth Furnace, and Thomas's family received Cornwall Furnace. When Thomas Burd died in 1837, his sons, William and Robert W. Coleman, managed Cornwall Furnace. The next generation of Colemans saw their fortune swell once again.

Cornwall Iron Furnace showed a $20,000 profit in 1840 and carried on with successful production of charcoal cold-blast iron in spite of increasing competition from anthracite blast fur-naces and other technical innovations. Improvements were made to the operation of the charcoal furnace in efforts to keep Cornwall Iron Furnace competitive with newer furnaces. In the late eighteenth century, blowing tubs had replaced the large, inefficient bellows. A steam engine arrived in about 1841 from the West Point Foundry in Cold Spring, New York, and it is likely that for a short time water and steam were used concurrently to provide power. A major remodeling project in the middle of the nineteenth century enlarged and rebuilt the furnace stack, from thirty or thirty-

Open-pit Mining at Cornwall Ore Mines, c. 1925. CORNWALL IRON FURNACE

fying it), mechanized cleanup methods, and use of a solid concrete arched lining, increased the efficiency and safety of this operation. The 1940 production amounted to 358,000 tons, more than had been taken in the first fifty years of mining; further improvements would soon raise this to 1.1 million tons a year.

Mining at Cornwall came to an abrupt end in 1973, after flooding from Tropical Storm Agnes filled the mine shafts. Mine engineers determined that the remaining quantity and quality of ore did not justify the expense of dewatering the mine. The open-pit mine is now a lake estimated to be four hundred feet deep.

The Cornwall Mines had seen an unparalleled 234 years of uninterrupted production, from Revolutionary times until the twentieth century, through four major wars. Total production was 106 million tons of iron ore.

one to thirty-two feet tall, reducing the bosh from nine to seven feet, and gave the structures their red sandstone Gothic Revival features. Cornwall Anthracite Furnace was eventually built by William and Robert W. Coleman in 1849 near the present-day community of Anthracite (Goosetown).

Considerable confusion arose among the principal mine owners, and in 1848 an agreement was drawn up stating that all owners were to account to one another for all ore removed. While the intent was honorable, it failed because the ore was found to differ in purity, and because some areas were easy to mine, whereas others required difficult, expensive mining methods. Ruinous competition among the many owners quickly developed into a dispute that ultimately led to one of the most famous and lengthy lawsuits in the history of the Lebanon County courts. Known as Alden's Appeal, it entered the courts on July 15, 1856, and, due to appeals, remained in the legal system until 1880.

In 1864, the Cornwall Ore Banks Company was formed by a compromise of the ninety-six part owners to maintain some stability in production, since the original lawsuit was still in the courts. J. Taylor Boyd was put in charge of production, and for the remainder of the century, this company was in control of all mining and responsible for accounting to the joint owners for all ore taken, with the notable exception of the independent Robesonia Iron Company.

Iron was in much demand during the War Between the States, but due to fluctuations in the tariff statutes, iron-masters in the North did not realize profits until 1864. During this period, in an almost exact reverse of the Revolutionary War experience, competition with British iron products was intense. In its last decade of operation, from 1873 to 1883, Cornwall Furnace operated at a loss. The stage was set for Robert Habersham Coleman, the fourth and final generation of Colemans, to take over the iron business at Cornwall and in Lebanon.

Brothers Robert W. and William Coleman had shared ownership since the death of their father, Thomas Burd. Robert W. never married, and his properties passed to his sisters, Anne Alden, Margaret Freeman, and Sarah Coleman. William's passed to his children, Anne and Robert Habersham. Robert Habersham Coleman took over managing control of the family's holdings in 1879. He had completed his education at Trinity College in Hartford, Connecticut, and was newly married when his business acumen was acutely challenged by newer processes of steel production, the use of anthracite and coke, the discovery of Lake Superior iron ore deposits, and the urbanization of factories to locate them near rail terminals. He became totally immersed in business after the tragic death of his young bride, Lillie, in 1881. To keep up with the times, he opened a series of modern anthracite furnaces and shut down Cornwall Iron Furnace for good on February 11, 1883.

By 1889, the estimate of Robert Habersham Coleman's wealth had grown from $7 million to $30 million. He married Edith Johnstone of Baltimore and had five children. Robert was generous with his wealth, donating money to his alma mater and college fraternity, as well as various churches. He built houses, schools, and a church for his workers, and created the summer colony of Mount Gretna in 1889 as a pleasure stop on his Cornwall and Lebanon Railroad. He garnered controlling interest in the Dime Savings Bank in Lebanon and modernized Coleman farm holdings. He bought a railroad construction company in Florida and a fifty-mile stretch of railroad there. Then, in an abrupt series of events, the Coleman fortune vanished. Robert lost two costly lawsuits in 1891, with staggering results, followed by the stock market panic of 1893. At the age of thirty-seven, Robert Habersham Coleman was ruined financially and in poor physical condition with tuberculosis. He moved his family to Saranac Lake in the beautiful Adirondack Mountains of New York, where he lived as a recluse until his death in 1930.

William C. Freeman, who had established the North Cornwall Anthracite Furnace in 1871, continued in the Cornwall businesses until his death in 1903. His only son, William Coleman Freeman, became a banker, a Pennsylvania state senator, and a state secretary of banking, whose active involvement in the iron business ended with the sale of Robesonia Iron Company to Bethlehem Steel Corporation. After 1883, holdings of the Cornwall Ore Bank Company had begun to pass into the hands of the Pennsylvania Steel Company, and in 1916, Bethlehem Steel Company had

TRANSPORTING IRON

Transporting iron ore, pigs, and iron products is no easy task, because iron is heavy. As local customers for iron products could not entirely support a furnace, expanding the market depended on access to transportation routes—roads or waterways. Over the years, Cornwall Furnace is known to have supplied pig iron to the Tulpehocken Forge in Berks County, presumably overland in wagons, and to New Cumberland Forge in Cumberland County, likely via the Susquehanna River. What was known as the "Great Pig Iron Road" ran from Lancaster to the mouth of the Susquehanna in Maryland, roughly paralleling U.S. Route 222. Iron from Cornwall was also sold at several distant markets in the colonies, including Virginia.

Canals and then railroads provided ways for Cornwall iron to be transported ever farther. The Union Canal was dug in the early 1820s to connect the Susquehanna River with the Schuylkill River. It carried a variety of people and products, not the least of which were iron ore, pig iron, and other iron products, until it was closed in 1885. A plank road, completed in 1826, for mule- and horse-drawn wagons

Restored Union Canal Tunnel.

led from the Union Canal five miles north to Cornwall. The Cornwall Railroad, built in 1855, connected Cornwall to the Union Canal and brought anthracite coal to the Coleman furnaces. The railroad gradually expanded so that the transportation of iron was not interrupted when the canals became obsolete. Not only did railroads provide transportation for iron ore and iron products, but by the 1830s, they began to create a demand for iron and steel themselves.

entered the Cornwall scene by purchasing the Pennsylvania Steel Company. Bethlehem acquired ownership of the open pit mine from William, the last of the Coleman heirs, by 1919, finalizing its control at Cornwall with the acquisition of the Robesonia Iron Company in 1926. Mining and the iron industry continued at Cornwall and in Lebanon until the second half of the twentieth century, under the direction of Bethlehem Steel.

The last Coleman to occupy the ironmaster's mansion at Cornwall was Margaret Coleman Freeman Buckingham, the great-granddaughter of Robert Coleman and daughter of Margaret and William Freeman. In 1932, assisted by her nephew, State Senator William Coleman Freeman, she presented Cornwall Iron Furnace to the Commonwealth of Pennsylvania for a public museum. Margaret Buckingham, in typical Coleman fashion, was closely associated with the local community and always expressed the desire to have the mansion put to some useful purpose after the Colemans passed from the scene. Her wish was fulfilled posthumously in 1949, when her splendid estate and seventy-four acres of land were sold by William Coleman Freeman for a mere $20,000 to a corporation of Methodist laymen as a retirement home for Methodist clergy. Over the years, the property has been developed into a private retirement community.

Visiting the Site

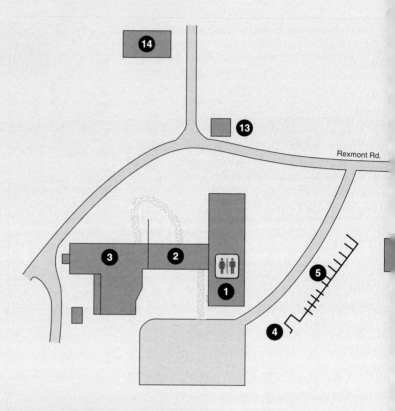

Rexmont Rd.

SITE LEGEND

1. Visitor Center
2. Connecting Shed
3. Furnace Building
4. Roasting Oven
5. Coal Bins
6. Blacksmith Shop
7. Wagon Shop
8. Abattoir
9. Stable*+

10. Manager's House and Mine Office*+
11. Open Pit Mine*+
12. Miners Village*+
13. Office Building*+
14. Ironmaster's Mansion*+

[restrooms icon] Restrooms

* Not open to the public
+ Not owned by the historic site

1 VISITOR CENTER

Located in the Charcoal House, the Visitor Center is the jumping-off place for tours and offers visitor amenities, including a reception area, museum store, and restrooms. Visits begin with orientation exhibits on mining, charcoal production, and ironmaking in general, as well as displays that tell the unique story of Cornwall. Little is known about this charcoal barn, whose four huge rooms, or bays, were once filled with charcoal—about one hundred days' supply, according to some records. Teamsters transported the finished charcoal from the colliers in the surrounding forests to the site, filling into each bay through its ground-level wall openings, now visible as windows. Baskets of charcoal were carried up the roof ladders and emptied into each bay through observable roof doors when the volume of charcoal precluded filling into the wall opening. All the bays of the Charcoal House empty into the second bay. From there, charcoal was taken to the furnace by way of the Connecting Shed.

② CONNECTING SHED

This roof, now reconstructed, was an important structure that protected the charcoal from inclement weather as the fillers pushed their buggies, or carts, from the Charcoal House to the furnace for charging. Here the visitor can sense the vast layout of the iron plantation, which once comprised almost ten thousand acres: the formidable Charcoal House facing the Furnace Building, the mine to the south, the Ironmaster's Mansion to the north, and surrounding villages, forests, and farmland that completed the plantation.

③ FURNACE BUILDING

The Furnace Building is actually several distinct buildings clustered around the furnace stack. Its elegant facade and Gothic Revival details were the result of a major mid-nineteenth-century remodeling project, and it stands as a testimony to the success of the furnace and to the refined tastes of its owners. The tour follows the path of the raw materials from the point of charging, or entering the stack, to casting the finished products.

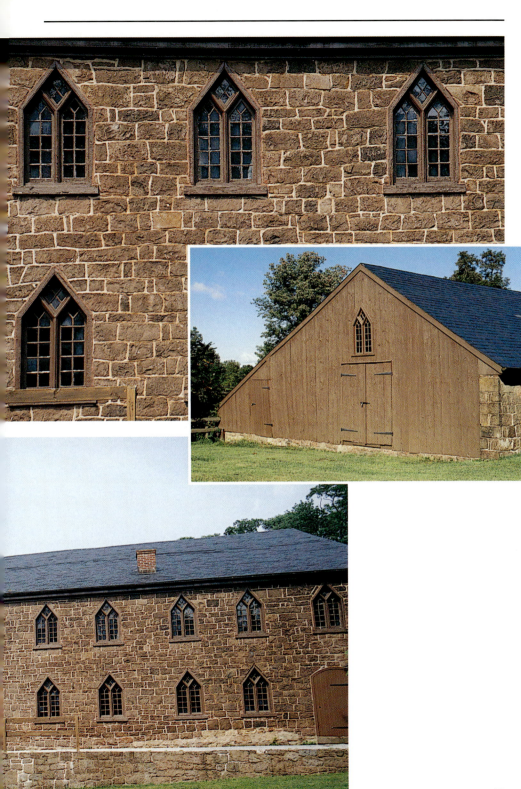

Charging Floor. Here the fillers, as directed by the founder, dumped their buggy loads of charcoal and iron ore, with the limestone, into the nineteen-inch mouth of the furnace, enduring relentless heat, flames, sparks, and soot. The larger buggy held the voluminous charcoal; the smaller buggy was used for the weighty iron ore. The shape of the furnace is clearly visible from its mouth, where visitors can observe the bosh, funnel, and crucible. Nineteenth-century founders were assisted by a call bell, a balance scale, and the steam boilers behind the mouth, which supplied a head of steam to the steam engine that powered the blast equipment. The tally board at the head of the stairs was used by fillers to count their buggy loads.

Wheel Room. In this room is housed the blast equipment believed to be the sole surviving example of this type of machinery intact from the period of its use. The original equipment housed here consisted of a water wheel that powered a large wood and leather bellows, which supplied the necessary blasts of air to the furnace. The more modern system consists of the great wheel turned by a gear on the crankshaft from the steam engine. As the wheel is turned, its axle turns with it, operating a reciprocal system of connecting rods and pistons in the blowing tubs. In this double manifold structure, the air is forced on both the up- and downstrokes into an equalizing chamber, or box, and from there by a pipe to the base of the stack. A great volume of air is produced, which is then compressed slightly

where it enters the stack through a four-foot cone called a tuyere (the Tuyere is seen later in the tour). This equipment could be operated year-round, unlike the water wheel, and its three tuyeres created a far more efficient blast than the bellows. The great wheel is twenty-four feet in diameter, weighs four tons, and turns on a thirteen-and-a-half foot wooden axle in the shape of a dodecagon. One source reports the axle making seven to nine revolutions per minute. The axle is the only piece of this equipment not original to the period of its use. It was reconstructed in order to demonstrate the operation of the entire system. The old axle is visible in the next room of the tour.

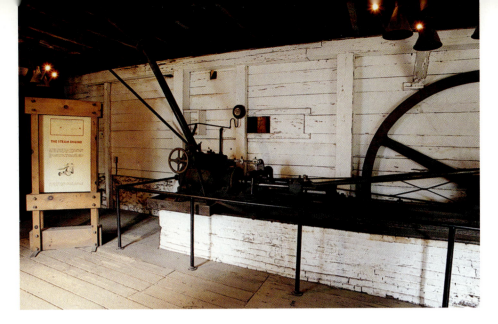

Engine Room. This steam engine, on its original housing, received a head of steam from the boilers behind the charging hole and powered the blast apparatus. A condenser retrieved and condensed the exhaust, which was then pumped back to the boilers, thus completing the circuit. From this location, a worker could monitor each part of the blast equipment, including the wheel, piston, and engine. If the blast equipment shut down for any reason, the charcoal would not burn hot enough to melt the rocks, a recipe for disaster in the furnace. This was such a strategic location that a second call bell (right) here permitted communication between the founder and the Engine Room. Thus the founder could signal the Engine Room to adjust the blast by modifying the engine speed in conjunction with the pressure gauge, and the Engine Room could signal the founder if there was trouble. In the last years of the furnace, it is said that one of the engine tenders, named James Shiner, was a cobbler who set up his bench in the corner of this room and mended shoes if everything was going well with the engine. The engine was cast at the West Point Foundry in Cold Spring, New York, and installed at Cornwall Furnace about 1841.

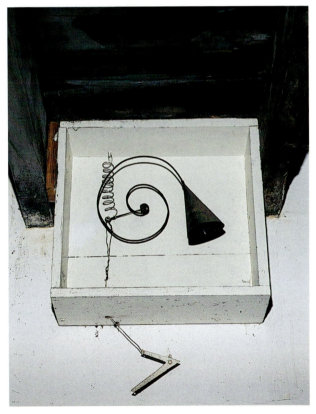

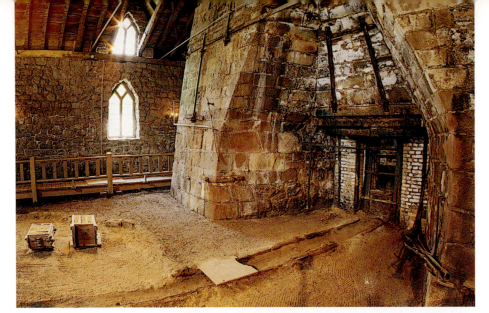

Casting Room. This area is also known as the Cast House or Casting Floor. The massive stone outer structure of the stack is visible from this location at the base of the furnace. The door to the crucible was lined inside with clay, and the circular and half-moon openings were plugged with clay. Twice a day the founder would tap the furnace, first releasing slag through the upper, round slag notch. The slag was diverted away from the casting sand floor, allowed to cool, and discarded. After removing the slag layer, the founder opened the half moon, and red hot molten iron flowed out of the furnace, guided by the gutterman through the "sow" and into the "pig" molds. Before tapping, the molder would have prepared the "pig bed," as well as any casting flasks for finished cast-iron products, such as sash weights, firebacks, skillets, grates, or stove plates. Working conditions in the casting room were quite harsh, and a great deal of skill was needed

by the molder, who was well paid for successful work. The pigs could be stored or transported to the forge, where they were refined into merchant bars of wrought iron. The cannon on display is one of forty-two such pieces cast for Revolutionary War Naval frigates. Little is known about the fate of the other cannons, but this one became defective on proving and remains as a valuable artifact.

Tuyere. This four-foot cone pressurized the blast from the blowing tubs before it entered the furnace. Most likely the Tuyere was angled up into the burning mass above the crucible, which is clearly visible through the Tuyere arch.

4 ROASTING OVEN

Alternate layers of charcoal and iron ore, loosely placed to admit the upward passage of air, were put into the Roasting Oven to effect the removal of sulfur from the iron ore. Failure to remove the sulfur caused difficulties in smelting. This structure was probably here by 1825, when the mine was beginning to yield a lower grade of ore.

5 COAL BINS

These bins are formed by walls that supported the railroad line into the mine. They were also used to store anthracite coal, which heated workers' houses and the Ironmaster's Mansion.

6 **BLACKSMITH SHOP**
The fabrication and repair of tools for mining and ironmaking were a constant concern. Here a blacksmith could perform these functions and make hardware for the community.

7 **WAGON SHOP**
Here wagons for the mining and ironmaking operations were built and repaired. Constructed sometime in the second half of the nineteenth century.

8 ABATTOIR

This charming Gothic building, featuring quatrefoil windows, served as a smokehouse and butcher shop for the Cornwall Estate. It was built in the nineteenth century and demonstrates the efforts to create a planned community in which all architecture was coordinated.

RELATED SITES

The following sites were part of the original Coleman plantation, but are not owned by the historic site. Visitors may observe these sites but should keep away from the entrances, as the sites are not open to the public.

9 STABLE

This building quartered the horses and mules used in everyday functions of the furnace, such as hauling raw materials and finished products.

10 MANAGER'S HOUSE AND MINE OFFICE

Present knowledge indicates that this impressive stone building was constructed in the nineteenth century as a residence for the furnace manager. Its size and design show the importance of the manager, who ranked second only to the owner. In the twentieth century, the building was enlarged and used for offices by Bethlehem Steel Corporation, owners of the Cornwall Ore Banks.

11 **OPEN PIT MINE**
The mine, now filled with water, is visible from Boyd Street, just south of Cornwall Iron Furnace. It operated continuously from the 1730s to 1973, was the largest open-pit iron ore mine in the world, and produced 106 million tons of iron ore, as well as copper, cobalt, gold, and silver.

12 **MINERS VILLAGE**
Just beyond the mine lies Miners Village, a picturesque community of lovely stone homes. The mine owners built this village a few units at a time to provide housing for workers.

13 **OFFICE BUILDING**

By 1875, this structure was an office serving the Cornwall Iron Company and Cornwall Estate. In the twentieth century, it served as the paymaster's office for Bethlehem Steel miners.

14 **IRONMASTER'S MANSION**

Curttis Grubb built this mansion in 1773 for his family after he assumed ownership of Cornwall Iron Furnace. It became the residence of William Coleman, son of Robert Coleman, in 1801 when he was appointed manager of the furnace. In 1865, Coleman engaged an Italian architect to remodel the home extensively, and the old Grubb home was transformed into a twenty-nine-room villa, which included a library, billiard room, two upstairs lounges, a spacious drawing room, an elegant dining room, a summer kitchen, and a tower overlooking a splendid estate. Full of interesting features, this home of elegant beauty was last occupied by the great-granddaughter of Robert Coleman, Margaret Coleman Freeman Buckingham. It is now part of a private retirement community.

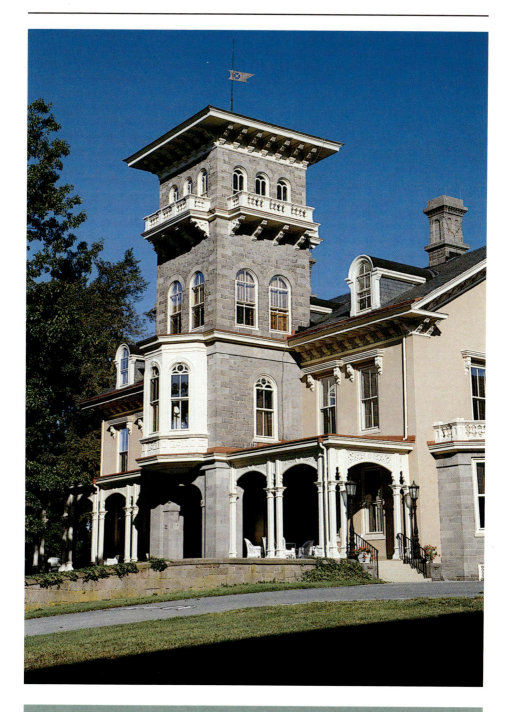

For information on hours, tours, programs, and activities at Cornwall Iron Furnace, visit **www.phmc.state.pa.us** or call **717-272-9711**.

Further Reading

Bining, Arthur Cecil. *Pennsylvania Iron Manufacture in the Eighteenth Century*. Harrisburg, Pa.: Pennsylvania Historical and Museum Commission, 1979.

Bomberger, Bruce, and William Sisson. *Made in Pennsylvania: An Overview of the Major Historical Industries of the Commonwealth*. Harrisburg, Pa.: Pennsylvania Historical and Museum Commission, 1991.

Dibert, James A. *Iron, Independence and Inheritance: The Story of Curttis and Peter Grubb*. Cornwall, Pa: Cornwall Iron Furnace Associates, 2000.

Eggert, Gerald G. *The Iron Industry in Pennsylvania*. University Park, Pa.: Pennsylvania Historical Association, 1994.

Gray, Carlyle, and Davis M. Lapham. *Guide to the Geology of Cornwall, PA*. Harrisburg, Pa.: Pennsylvania Geological Survey, 1961.

Kemper, Jackson. *American Charcoal Making in the Era of the Cold-blast Furnace*. 2nd. ed. Hopewell Furnace National Historic Site, Pa.: Eastern National Park and Monument Association, 1987.

Lapham, Davis, and Carlyle Gray. *Geology and Origin of the Triassic Magnetite Deposit and Diabase at Cornwall, Pennsylvania*. Harrisburg, Pa.: Commonwealth of Pennsylvania Department of Environmental Resources, 1973.

Lebanon's Royal Family. Lebanon, Pa.: Lebanon County Historical Society, 1996.

Miller, Frederic K. *The Rise of an Iron Community: An Economic History of Lebanon County, Pennsylvania from 1740–1865*. Lebanon, Pa.: Lebanon County Historical Society, 1950.

Noble, Richard E. *The Touch of Time: Robert Habersham Coleman, 1856–1930*. Lebanon, Pa.: Lebanon County Historical Society, 1983.

Oblinger, Carl. *Cornwall: The People and Culture of an Industrial Camelot, 1890–1980*. Harrisburg, Pa.: Pennsylvania Historical and Museum Commission, 1984.

Reed, Diane B. "The Magic of Mount Gretna: An Interview with Jack Bitner." *Pennsylvania Heritage* 28, no. 2 (spring 1992):16–23.

Silverman, Sharon Hernes. "A Blast from the Past; Cornwall Iron Furnace." *Pennsylvania Heritage* 24, no. 2 (spring 1998):20–31.

Also Available

Anthracite Heritage Museum
and Scranton Iron Furnaces

Brandywine Battlefield Park

Conrad Weiser Homestead

Daniel Boone Homestead

Drake Well Museum and Park

Eckley Miners' Village

Ephrata Cloister

Erie Maritime Museum and
U.S. Brig Niagara

Hope Lodge and Mather Mill

Joseph Priestley House

Landis Valley Museum

Old Economy Village

Pennsbury Manor

Railroad Museum of Pennsylvania

All titles are $10, plus shipping,
from Stackpole Books, 800-732-3669, www.stackpolebooks.com, or
The Pennsylvania Historical and Museum Commission, 800-747-7790,
www.phmc.state.pa.us